AS Biology
UNIT 2

Specification B

Module 2: Genes and Genetic Engineering

Keith Hirst

Philip Allan Updates
Market Place
Deddington
Oxfordshire
OX15 0SE

tel: 01869 338652
fax: 01869 337590
e-mail: sales@philipallan.co.uk
www.philipallan.co.uk

This Guide has been written specifically to support students preparing for the AQA Specification B AS Biology Unit 2 examination. The content has been neither approved nor endorsed by AQA and remains the sole responsibility of the author.

Typeset by Magnet Harlequin, Oxford
Printed by Information Press, Eynsham, Oxford

Contents

Introduction

■ ■ ■

Content Guidance

■ ■ ■

Questions and Answers

Introduction

About this guide

This guide is for students following the AQA Specification B AS Biology course. It deals with Unit 2, which examines the content of **Module 2: Genes and Genetic Engineering**. The key to success is examination technique. You should always have at the back of your mind the type of questions that can be asked, when both learning and revising a topic. This Introduction is devoted to the aims of the specification and to learning and revision skills. The Content Guidance section provides an outline of the topics you need to know and understand and includes detailed explanations of some of the topics. The Question and Answer section contains sample unit test questions, together with candidate responses which are accompanied by examiner's comments.

The best way to use this book is to:

- revise a topic using the Content Guidance section as a guide
- attempt the relevant question(s) *without looking* at the candidate responses
- compare your responses with the candidate responses and examiner's comments and see what marks you might have achieved
- revise the parts of the topic for which you did not obtain high marks

The aims of the AS specification

AS biology encourages you to:

- develop knowledge and understanding of concepts of biology
- develop the skills to use this knowledge and understanding in new situations
- develop an understanding of the methods used by scientists
- be aware of advances in technology that are relevant to biology
- recognise the value and responsible use of biology in society
- sustain and develop an interest in, and enjoyment of, biology

How the unit test assesses these aims

Very few marks in the unit test are given for simple recall of knowledge. Most of the marks are given for being able to:

- demonstrate understanding of concepts
- apply knowledge and understanding

These two areas include many skills, most or all of which will be addressed later in the unit test. In summary, you should be able to do the following.

- Draw on your knowledge to show understanding of the ethical, social, economic, environmental and technological implications and applications of biology. Read scientific articles in newspapers and periodicals, and watch documentaries on current affairs that deal with scientific issues so that you are aware of different viewpoints on controversial issues.
- Select, organise and present relevant information clearly and logically.
- Practise answering the longer section B questions that involve writing continuous prose.
- Describe, explain and interpret phenomena and effects in terms of biological principles and concepts, presenting arguments and ideas clearly and logically. Make sure you know the difference between the 'trigger' words *explain* and *describe*.
- Interpret (and translate from one form into another) data presented as continuous prose, or in tables, diagrams, drawings and graphs. You will be presented with data in many different forms in the unit test. Make sure that you have practised questions involving comprehension, graphs, tables and diagrams.
- Apply biological principles and concepts in solving problems in unfamiliar situations including those relating to the ethical, social, economic and technological implications and applications of biology. There will always be unfamiliar data in the unit test and the examiner will ask you to 'suggest' explanations. Again, make sure you have practised many examples of this type of question.
- Assess the validity of biological information, experiments, inferences and statements. Don't leave your experimental skills in the laboratory. Interpreting and evaluating data questions will usually appear in the unit tests.

A unit test will comprise questions that require the use of most, if not all, of these skills.

Weightings

Unit 2 is assessed by a unit test and carries 30% of the marks for the AS. Of this 30%:

- 18% of the marks are given for demonstrating knowledge and understanding of the unit content
- 12% of the marks are given for being able to apply this knowledge and understanding in new situations

Command terms

Examiners use *trigger words* to advise you which skill they are testing. You must know what the examiner wants when these trigger words appear in a question.

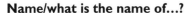

Name/what is the name of...?

This usually requires a technical term or its equivalent. Answers to this type of question normally involve no more than one or two words. Do not waste time by repeating the question in the answer.

List...

This requires you to give a number of features or points, each often no more than a single word, so do not go into further detail.

Define/what is meant by...?

'Define' requires a statement giving the meaning of a particular term or word. 'What is meant by...?' is used frequently in questions on a comprehension passage. It emphasises that a formal definition as such is not required.

Outline...

This means give a brief summary of the main points. There are two good indications as to the amount of detail required. These are the mark allocations and the space allowed for the answer — usually two lines per mark.

Describe...

This means no more than it says: 'Give a description of...'. So 'Describe a curve on a graph' requires a description of the shape of the curve, preferably related to key points or values; 'Describe an experiment' means give an account of how such an experiment might be carried out.

Describe how you...

The emphasis here is on the word *you* and the expression is often used when asking questions about experimental design. What is required is an account of how something could be done by you as a student working in an ordinary school or college laboratory.

Evaluate...

Evaluating is more than just listing advantages and disadvantages. It requires an explanation. Evaluating the evidence for and against a particular point of view requires an explanation of each of the points being made.

Explain...

This requires you to give a reason or interpretation, *not* a description. The term 'describe' answers the question 'what?'. The term 'explain' answers the question 'why?'. Thus, 'Explain a curve on a graph' requires a biological reason for any change of direction or pattern that is evident.

Suggest...

Suggest is used when it is not possible to give the answer directly from the facts you have learned. The answer should be based on your general understanding of biology rather than on recall of learnt material. It also indicates that there may be a number of correct alternatives.

Give the evidence for... /using examples from...

Answers to questions involving these phrases must follow the instructions. Marks are *only* awarded for appropriate references to the information provided in the question.

Plot/sketch...

These terms refer to the drawing of graphs. 'Plot' means that the data should be presented as an appropriate graph on graph paper with the points plotted accurately. 'Sketch' requires a simple estimate of the expected curve, and can be made on ordinary lined paper. However, even in a sketched graph, the axes should be correctly labelled.

Calculate...

This term is used where the only requirement is a numerical answer expressed in appropriate units. The additional instruction, 'Show your working', will be used if details or methods are required. Make sure that you can calculate percentages and proportions, since these appear in most unit tests.

Revision planning

Key words

A biological specification contains so many unfamiliar words it can appear to be a foreign language. It is important that you know the meaning of all of these words so that you know what is being asked in a question and can use the words correctly in your responses.

Below are some extracts from Module 2. The biological words that you need to know are in bold.

- **Genes** are sections of **DNA**, which contain coded information that determines the nature and development of organisms.
- A gene can exist in different forms called **alleles**, which are positioned, in the same relative position (**locus**) on **homologous chromosomes**.
- DNA is a stable **polynucleotide**.
- The double-helix structure of the DNA molecule in terms of:
  - the components of DNA nucleotides
  - the sugar–phosphate backbone
  - specific **base pairing** and **hydrogen bonding** between polynucleotide strands

It is a good idea to go through the specification content listed in Content Guidance, underlining biological words and then writing a definition of each one, for example:

Homologous chromosomes: a pair of chromosomes with the same sequence of genes. Occur in a diploid nucleus.

Nucleotide: one of the monomers that makes up a nucleic acid. Contains a sugar, a phosphate group and a base.

This will give you the biology vocabulary that is essential both to understand and to answer questions.

Revision progress

You might find it useful to keep track of how your revision is going by drawing the table below, including the topics in the first column.

Module topic	Revised (N/P/F)	Self-evaluation (1–5)
The genetic code	F	5
The cell cycle	F	5
Sexual reproduction and fertilisation	P	3
Applications of gene technology	P	2

Complete column 2 to show how far you have got with your revision.
   N = not yet revised
   P = partly revised
   F = fully revised

Complete column 3 to show how confident you are with the topic.
   5 = I am confident I could answer any question on this topic
   1 = I found the practice questions very difficult.

Update the table as your revision progresses.

Revising at home

- Revise regularly — do *not* leave revision until near the examination.
- Plan your revision carefully so that there is no last-minute rush.
- Revise in a quiet room — you cannot revise properly if distracted by the television or music.
- Revise in short stretches — work for half an hour, have a breather for 10 minutes, then start again. You should be able to revise for about 2–3 hours in an evening.
- Revise actively — read a topic, then close your book and make a summary from memory. Then go back and see what you've missed.
- Do as many questions as possible from sample and past papers.

In the exam room

- Think before you write.
- Don't waste time copying out the question.
- Make a plan for longer answers.
- Think in paragraphs.
- Don't rush.
- Don't panic — if you can't do a question, go onto the next one.
- Check your spelling of words that are similar to others.

Content
Guidance

This section provides an overview of the key terms and concepts covered in **Module 2: Genes and Genetic Engineering**. The major facts that you need to learn are outlined and the principles you need to understand are explained. Some AS questions will test recall while others will test understanding. For example, you must know the principles involved in the translation and transcription of the genetic code of DNA, but you will also be given examples of different codes and asked to translate and/or transcribe them. A third type of question involves application. For example, you will be given data about the life cycle of an organism that you have not studied and asked to interpret them. Again, you can only do this if you understand the principles involved in the formation of body cells and gametes.

The content of Module 2 falls into four main areas:

(1) The genetic code
This topic looks at the structure of nucleic acids and at DNA replication. The role of nucleic acids in protein synthesis and the effects of mutations are also considered.

(2) The cell cycle
This involves a detailed study of mitosis, followed by a look at applications of mitosis in cloning.

(3) Sexual reproduction and fertilisation
This topic looks at the structure of gametes, and the role of meiosis and mitosis in life cycles.

(4) Applications of gene technology
This looks at the basic techniques used in genetic engineering. It then considers the ways in which these techniques are used in producing chemicals, treating disease and producing organisms with new characteristics.

The genetic code

Nucleic acids

Nucleic acids are polymers made up from units called **nucleotides**. Each nucleotide has the same basic structure — a 5-carbon sugar to which are attached a phosphate group and a nitrogen-containing base.

Phosphate ◯

Five-carbon sugar

Nitrogen-containing base

DNA

The **DNA molecule** (deoxyribonucleic acid) consists of two polynucleotide strands twisted together. They are held in place by hydrogen bonds, forming a double helix. One strand, known as the sense strand, contains the genetic code.

The sugar molecule in DNA nucleotides is deoxyribose. The nitrogenous base in each nucleotide can be adenine, cytosine, guanine or thymine. These form complementary base pairs:
- an **adenine** base of a nucleotide from one chain always pairs with a **thymine** base of a nucleotide in the opposite chain (A–T);
- a **cytosine** base of a nucleotide from one chain always pairs with a **guanine** base of a nucleotide in the opposite chain (C–G).

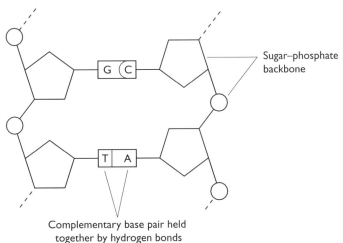

Sugar–phosphate backbone

Complementary base pair held together by hydrogen bonds

DNA replication

Before the DNA molecule can replicate, it must separate into its two constituent strands. The hydrogen bonds that hold the two strands together are broken by an enzyme. A second enzyme, **DNA polymerase**, is needed to join free nucleotides together to form a complementary strand for each of the original DNA strands. The original strands act as templates for the production of the new strands. Each of the new nucleotides added has a base that is complementary to a base in the original DNA strands. Thus the two new DNA molecules contain exactly the same sequence of nucleotides as the original strands.

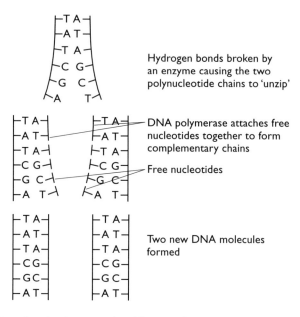

Hydrogen bonds broken by an enzyme causing the two polynucleotide chains to 'unzip'

DNA polymerase attaches free nucleotides together to form complementary chains

Free nucleotides

Two new DNA molecules formed

The two new DNA molecules have nucleotide strands with exactly the same sequence of bases as the original molecule. Half of each new molecule comes from the original molecule, so this method of replication is known as **semi-conservative replication**.

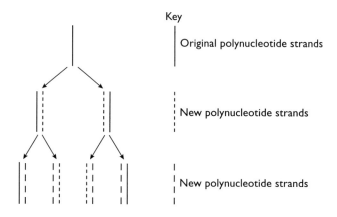

Key

Original polynucleotide strands

New polynucleotide strands

New polynucleotide strands

Structure in relation to function

The sequence of bases in DNA enables it to store information. The information is in the form of **codons**. Each codon is a sequence of three nucleotide bases. Some codons code for specific amino acids, others code for 'punctuation marks' e.g. 'start here' or 'stop here'.

The extreme length of the molecule means that large amounts of information can be stored. This information can be replicated exactly. The bonding of the sense strand to the opposite strand by hydrogen bonds makes DNA a very stable molecule, maintaining the integrity of the information.

RNA

RNA molecules differ from DNA molecules in three important ways:
- the molecules are single-stranded rather than double-stranded
- the sugar molecule is ribose rather than deoxyribose
- the base thymine is replaced by the base uracil

Messenger RNA is a copy of part of the genetic code of DNA. The process of copying the DNA code into RNA code is called **transcription**.

DNA molecule in nucleus unzips

RNA polymerase joins RNA nucleotides together to form messenger RNA molecule

Messenger RNA molecule leaves nucleus and becomes attached to ribosomes

The function of messenger RNA molecules is to carry part of the DNA code to the ribosomes where the code will be used to produce protein molecules. Messenger RNA molecules are much shorter than DNA molecules, so they are able to pass through the pores of the nucleus en route to the ribosomes.

The code on the mRNA is now **translated**. This is done by **transfer RNA (tRNA)** molecules. A tRNA molecule has a 'clover-leaf' shape.

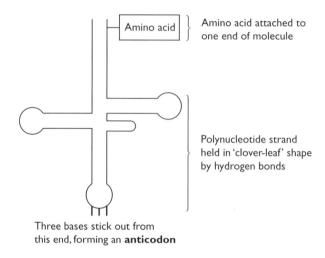

Three bases stick out from
this end, forming an **anticodon**

There are 20 different amino acids and each of these will only attach to a specific tRNA molecule. Thus there are 20 different tRNA molecules. The amino acid becomes attached to one end of a tRNA molecule. At the other end of the tRNA molecule is a sequence of three nucleotides. The bases of these three nucleotides form an **anticodon**. This anticodon has bases that are complementary to the mRNA base code for the amino acid — the **codon**. tRNA molecules bring their specific amino acids to the mRNA chain. Since the mRNA codons are specific to particular tRNA anticodons, the amino acids are joined together in the correct sequence during protein synthesis.

Protein synthesis

The following diagram shows the main stages in protein synthesis.

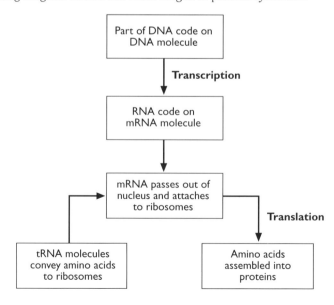

Be careful not to confuse transcription and translation. **Transcription** is the copying of part of the DNA code to messenger RNA code. **Translation** is using the mRNA code to assemble amino acids in the correct order to produce the specific protein. Transcription occurs in the nucleus; translation occurs at the **ribosomes**.

The code on the mRNA molecules is in the form of a **triplet codon** copied from the DNA. Anticodons on the tRNA molecules bind to the complementary codons on the mRNA molecules. The amino acids attached to the tRNA molecules are then joined to form polypeptide molecules, as shown below. As they are formed, the polypeptide chains join and/or fold and/or twist to form protein molecules.

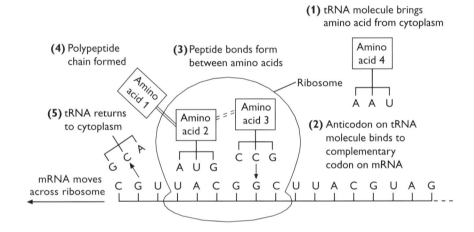

Genes are sections of DNA which contain coded information. Many genes contain information to produce a specific protein. A gene can exist in different forms called alleles which are positioned in the same relative position (locus) on homologous chromosomes. The different alleles of a gene may code for different proteins.

Mutation

A **mutation** is a change in the genetic code. New forms of alleles arise from changes (mutations) in existing alleles. Mutations occur naturally at random. High-energy radiation, high-energy particles and some chemicals are mutagenic agents.

A gene mutation is the result of a change in the sequence of bases in DNA. There are three ways in which the sequence of bases can be changed: **addition**, **deletion** or **substitution**.

- In addition, an extra base (or bases) is attached to the sequence. For example, if T is added between G and C to AAGCTACAC the sequence becomes AAGTCTACAC.
- In deletion, a base (or bases) is deleted from a sequence. For example, if T is deleted from AAGCTACAC the sequence becomes AAGCACAC.
- In substitution, a base (or bases) is substituted in a sequence. For example, if T is substituted by A in the sequence AAGCTACAC the sequence becomes AAGCAACAC.

The change in the DNA sequence can result in a change in the sequence of amino acids that are added to a protein. For example, in the addition mutation described above the original triplets are:

AAG   CTA   CAC

When T is added, the triplets become:

AAG   TCT   ACA

TCT would code for a different amino acid than would CTA. Similarly, ACA would code for a different amino acid than would CAC. Thus, part of the protein molecule formed would be different. If the protein molecule was an enzyme, the change in amino acid sequence might affect the tertiary structure and thus the active site. The enzyme would then be unable to bind to the substrate and hence would not work.

If the enzyme was one of a number in a metabolic chain, for example Enzyme B in the chain

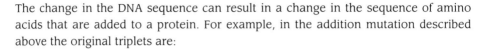

then Substrate 3 would not be formed and consequently the Product 4 would not be produced.

The cell cycle

Mitosis

During mitosis DNA *replicates* in the parent cell, which divides to produce two new cells, each containing an *exact copy* of the DNA of the parent cell. Mitosis *increases cell number* in this way in growth and tissue repair, asexual reproduction and vegetative propagation.

You need to be able to recognise drawings or photographs of the stages in mitosis and to explain what is happening in each. These are summarised below.

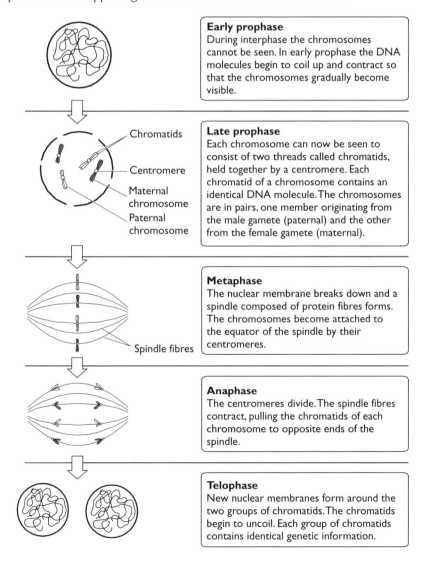

Early prophase
During interphase the chromosomes cannot be seen. In early prophase the DNA molecules begin to coil up and contract so that the chromosomes gradually become visible.

Chromatids
Centromere
Maternal chromosome
Paternal chromosome

Late prophase
Each chromosome can now be seen to consist of two threads called chromatids, held together by a centromere. Each chromatid of a chromosome contains an identical DNA molecule. The chromosomes are in pairs, one member originating from the male gamete (paternal) and the other from the female gamete (maternal).

Metaphase
The nuclear membrane breaks down and a spindle composed of protein fibres forms. The chromosomes become attached to the equator of the spindle by their centromeres.

Spindle fibres

Anaphase
The centromeres divide. The spindle fibres contract, pulling the chromatids of each chromosome to opposite ends of the spindle.

Telophase
New nuclear membranes form around the two groups of chromatids. The chromatids begin to uncoil. Each group of chromatids contains identical genetic information.

The cell cycle

Mitosis takes up only a small proportion of the total time required for one complete cell cycle. The rest of the time is taken up with growth, DNA replication and protein synthesis.

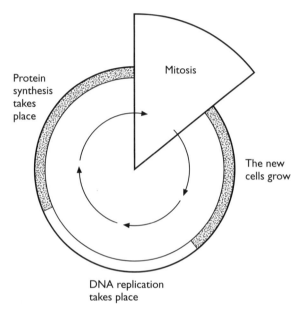

Cloning

Genetically identical organisms (clones) can be produced by:
- vegetative propagation
- splitting embryos

In both these processes, all the new cells are produced by mitosis so the offspring have identical genetic information to the parent.

Vegetative propagation

No specific examples are given in the specification, so you should be *aware* (but have no need to know in detail) that there are several different ways of producing clones of plants. These include:
- cuttings, e.g. chrysanthemum
- tubers, e.g. potato
- bulbs, e.g. daffodil
- corms, e.g. crocus
- rhizomes, e.g. iris
- runners, e.g. strawberry
- budding, e.g. rose
- grafting, e.g. apple

Some examples of types of vegetative propagation are illustrated below.

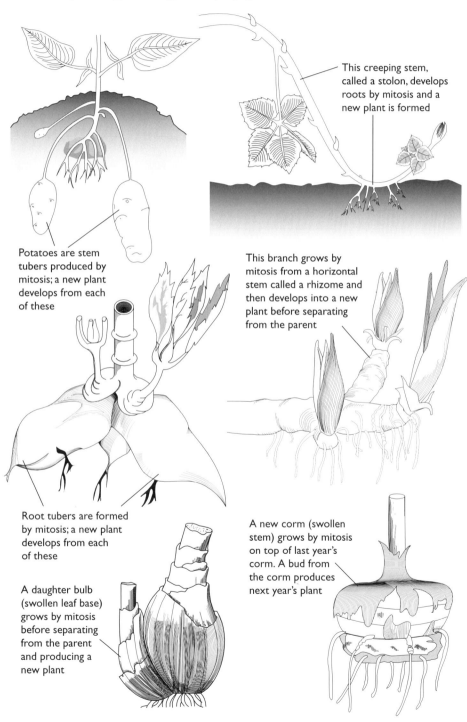

This creeping stem, called a stolon, develops roots by mitosis and a new plant is formed

Potatoes are stem tubers produced by mitosis; a new plant develops from each of these

This branch grows by mitosis from a horizontal stem called a rhizome and then develops into a new plant before separating from the parent

Root tubers are formed by mitosis; a new plant develops from each of these

A daughter bulb (swollen leaf base) grows by mitosis before separating from the parent and producing a new plant

A new corm (swollen stem) grows by mitosis on top of last year's corm. A bud from the corm produces next year's plant

In each case, part of a plant is separated from the parent and then either grown independently or joined onto another plant.

Cloning animals

Clones of animals can be obtained by separating the cells of developing embryos. This procedure is summarised in the following diagram.

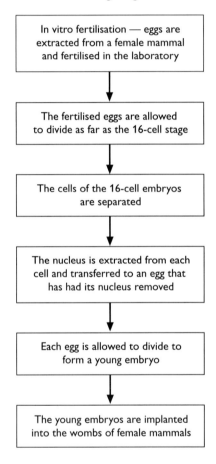

Each egg that has had its nucleus replaced contains identical genetic information, so all the embryos will be clones of the original embryo.

This technique is used to produce clones of farm animals such as cattle and sheep. The embryos produced have identical genetic information, since all cell divisions of the zygote are by mitosis. They will, however, contain alleles from both male and female parents, so the embryos will have genetic information from each of the parents that is slightly different.

'Dolly' the sheep was produced in a different way. A cell was removed from the udder of one adult female sheep. An unfertilised egg was taken from another adult sheep and the nucleus removed. The nucleus of the egg was then replaced by the nucleus from the udder cell. This egg was stimulated to divide and implanted into the womb of a ewe. The embryos produced in this way all had genetic information *identical* to that of the adult from which the udder cell was taken.

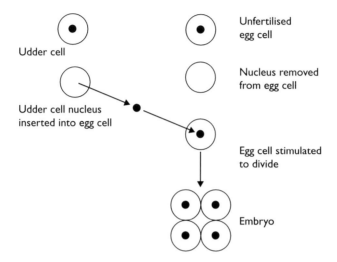

Sexual reproduction and fertilisation

Gametes and fertilisation

In sexual reproduction, DNA from one generation is passed to the next generation by gametes.

Sexual reproduction involves two stages:
- gamete formation (e.g. eggs and sperm)
- fertilisation (e.g. fusion of an egg and a sperm)

Female gametes are usually larger than male gametes since they contain food for the developing embryo. Mammalian eggs are small since they need only to contain enough food for the early stages of the development of the embryo; for the rest of the development, food is supplied from the mother via the placenta.

Male gametes are usually motile while female gametes are usually non-motile. In animals the sperm usually swim to reach the female gamete. Most sperm have tails to enable them to swim.

In flowering plants the male gametes are nuclei. These nuclei are transported from the male organ, the anther, to the female organs inside pollen grains. The pollen grains themselves are usually transported from anther to stigma by either wind or insects. After landing on the stigma, the pollen grains germinate, forming pollen tubes that grow towards the ovary. The male nuclei travel down these tubes and fuse with the female nuclei inside the ovary.

Meiosis

During **meiosis**, cells containing pairs of homologous chromosomes divide to produce gametes containing *one* chromosome from *each* **homologous pair**.

In meiosis the number of chromosomes is reduced from the **diploid** number ($2n$) to the **haploid** number (n). In humans, body cells have 46 chromosomes (23 pairs). The gametes (eggs and sperm) therefore have 23 chromosomes.

Fertilisation

When gametes fuse at fertilisation to form a **zygote** the diploid number is restored. The chromosomes from male and female gametes restore the homologous pairs.

This enables a constant chromosome number to be maintained from generation to generation. (Otherwise, if human gametes had 46 chromosomes, the next generation would have 92 chromosomes, the next 184 and so on.)

Life cycles

The life cycle of a mammal is best explained as a flow chart.

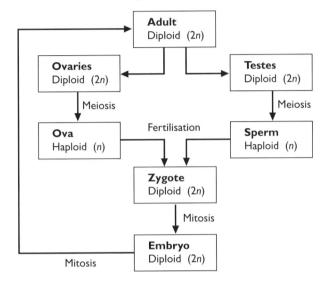

Other organisms have life cycles in which some organisms are diploid and others haploid. No life cycles are specified by the specification, so there will be questions where you will have to work out which organisms are haploid and which are diploid. Remember that gametes are always haploid, but that they can be formed by meiosis in some organisms and by mitosis in others.

The following diagram gives an example of a life cycle where gametes are formed by mitosis.

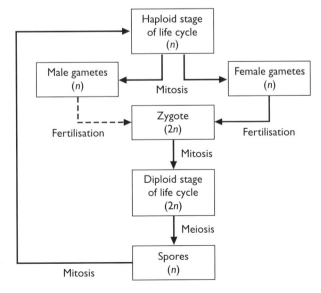

This type of life cycle is found in ferns. The fern plant you see growing in a wood is the **sporophyte** — the spore-producing plant. The fern plant is diploid. The spores are produced by meiosis and are therefore haploid. When they land on damp soil the spores germinate to produce a flat structure about 1 cm in diameter. This is the **gametophyte** — the gamete-producing stage. The gametophyte is haploid. The gametes are produced by mitosis. The male gametes have whip-like flagellae, which they use to swim towards the female gamete for fertilisation. The zygote is diploid and it develops into the diploid sporophyte.

Applications of gene technology

Basic principles

In genetic engineering, genes are taken from one organism and inserted into another.

- The cells have first to be broken open and the nuclear membrane disrupted to remove the DNA.
- An enzyme called a **restriction endonuclease** is used to extract the relevant section of DNA.
- The removed piece of DNA has 'sticky ends', each consisting of a single strand with a few base pairs.
- The same restriction enzyme is used to cut the DNA from a vector.
- The vector DNA will have sticky ends with bases complementary to those of the DNA to be inserted.
- The vector is usually a bacterial plasmid or a virus.
- The sticky ends of the two pieces of DNA are joined together using a **ligase enzyme**.
- The vector that now contains **recombinant DNA** is then introduced into the cells of the target organism.

This procedure is summarised below.

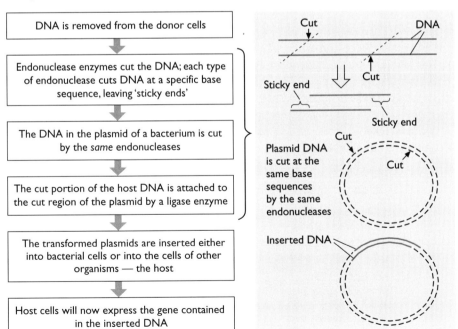

Genetically engineered microbes

Rapid reproduction of microorganisms enables a transferred gene to be cloned, producing many copies of the gene. Bacteria containing the transferred gene can be cultured on a large scale in industrial fermenters. Useful substances produced by using genetically engineered microorganisms include antibiotics, hormones and enzymes.

Plasmids containing the recombinant DNA are difficult to introduce into bacteria. The technique is to 'heat shock' the recipient bacteria by first placing them into cold calcium chloride solution, and then warming them up. This alters the properties of the cell membrane. Recombinant DNA plasmids enter *some* of the cells.

The problem is to identify which bacteria have recombinant DNA plasmids in them. To do this, a second gene is added to the recombinant DNA plasmids. This is usually a gene for resistance to a specific antibiotic. If one of these plasmids enters a bacterium, the bacterium will then contain both the required gene and the gene conferring antibiotic resistance. If the bacteria are grown in a medium containing the specific antibiotic, only bacteria with *both* genes will grow.

The polymerase chain reaction

The process of DNA replication can be made to occur artificially and repeatedly in a laboratory process called the **polymerase chain reaction** (PCR):
- The DNA sample is heated to a temperature of 95 °C. This separates it into its two strands.
- The sample is mixed with DNA nucleotides, DNA polymerase and a primer, which is a short piece of DNA.
- The primer directs the enzyme to where it is to start copying.
- The temperature is reduced to 40 °C. The primer will bind to the piece of DNA which is being copied.
- The temperature is raised to 70 °C. DNA polymerase now copies each strand of the original DNA. At the end of this part of the reaction there will be two molecules of DNA, each identical to the original one.
- This cycle can then be repeated, each time doubling the amount of DNA.

This procedure is summarised below.

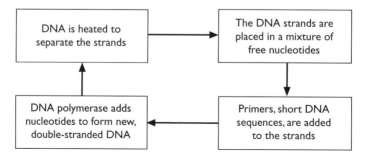

DNA sequencing

DNA sequencing is finding the sequence of nucleotides in a DNA sample. The procedure is as follows:

- DNA is isolated from a suitable sample.
- The DNA is cut into smaller pieces using restriction enzymes. Some of these pieces will contain the repeated sequences that are being investigated.
- The PCR reaction is used to multiply these fragments.
- Electrophoresis is used to separate the pieces. The DNA is placed on a layer of gel and an electric current applied. The DNA pieces separate out with the smaller pieces travelling further than the larger ones.
- The end result is that there are bands of pieces of DNA of different lengths.
- These are transferred to a sheet of nitrocellulose.
- Radioactive probes are now used to identify the pieces of DNA. These are single-stranded pieces of DNA with complementary base pairing for a specific base sequence. The probes attach to complementary DNA pieces.
- The sheet of nitrocellulose is then placed next to unexposed photographic film. The radioactivity in the probes shows up as dark bands on the film.

One application of this is in forensic science, producing genetic fingerprints. If the sequence of nucleotides in a forensic sample of DNA is found to be the same as that in a sample of DNA from the suspect, then there is a high likelihood that the forensic sample came from the suspect.

In the Human Genome Project, the sequence of nucleotides has been determined for all the DNA of a sample of humans. Scientists are now determining which parts of the sequence are particular genes. When the genes have been identified there is the possibility of using gene technology to cure diseases caused by mutations of these genes.

Gene therapy

Some types of genetic disease result from mutation of a gene responsible for producing a particular protein in the body. The protein could be an enzyme, or it could be a structural protein, such as a channel protein in a cell membrane. Scientists are developing methods of delivering the healthy gene for the protein to the part of the body affected by the disease so that the cells in that organ will produce the healthy version of the protein. This technique has already been successfully employed in treating genetic diseases of the lungs and of the liver.

Cystic fibrosis

In cystic fibrosis a transmembrane regulator protein, CFTR, is defective. This is because one amino acid is missing from the polypeptide. The function of this gene is to transport chloride ions through the cell membrane. In the lungs, cells with the defective protein retain chloride ions and this causes them to retain water. The mucus layer lining the lungs receives less water and so becomes thicker. The cilia in the lungs that normally remove mucus are unable to remove this thick mucus so it accumulates in the lungs. This means that less oxygen is absorbed into the blood via the lungs.

Mucus also accumulates in the pancreatic duct. This prevents pancreatic enzymes reaching the food in the digestive system.

A mutant of the gene that produces CFTR results in CFTR with one missing amino acid. In gene therapy healthy genes may be cloned and used to replace defective genes. Techniques that could be used to introduce healthy CFTR genes into lung epithelial cells include:

- use of a harmless virus into which the CFTR gene has been inserted. Viruses inject their own DNA into cells. This DNA takes over from the cell's own DNA. If the virus DNA contains the gene for normal CFTR the cells will manufacture normal CFTR. The virus is made harmless by removing the genes that cause damage to human cells, but leaving intact the genes that produce proteins used to insert genetic information into human cells.
- wrapping the gene in lipid molecules that can pass through the membranes of lung cells.

Genetically engineered animals

Some diseases are caused by the inability of the patient to produce a particular protein. These diseases can sometimes be treated by giving the patient synthetic protein. Proteins are extremely complicated molecules and therefore difficult to synthesise in the laboratory, so gene technology is used to make animals such as sheep produce the protein. Using the techniques outlined below, the healthy gene is extracted from human cells and inserted into fertilised sheep eggs. Another gene called a promoter gene is also inserted into the eggs. When the egg has developed into an adult sheep, this promoter gene stimulates the udder cells of the sheep to synthesise the protein along with the other components of the milk. The protein can then be harvested from the milk and given to the patient.

Animals that have been genetically modified in this way are called transgenic animals. Since all the cells of a transgenic animal contain the human gene, the animal transmits the gene to its offspring.

AAT

Alpha-1-antitrypsin (AAT) is a protein used to treat emphysema and cystic fibrosis. It is very difficult to obtain this chemical from human blood. However, sheep have been genetically engineered to produce AAT.

- The AAT gene is extracted from human cells using restriction endonuclease enzyme and inserted into bacterial plasmids using ligase enzyme.
- Eggs are removed from a sheep and fertilised in vitro.
- The plasmids containing the AAT are injected into the fertilised egg with a micropipette.

- The egg is allowed to divide to the 16-cell stage and then implanted into the womb of a female sheep.
- Sheep produced in this way have milk that contains large amounts of AAT.

Issues

It is worthwhile collecting articles from newspapers, magazines and the websites of pressure groups such as Greenpeace that deal with issues concerned with the applications of gene technology. These articles will most often contain views against the use of gene technology. You should be able to summarise these views as well as the benefits of using these techniques. Then you will be able to evaluate the use of a particular technique.

One example might be the introduction of a gene for herbicide resistance into a crop plant. The crop could then be sprayed with herbicide that would kill the weeds but not affect the crop. This would reduce competition for nutrients and therefore increase the crop yield, bringing more profit for the farmer.

However, there is a possibility of gene transfer from the crop to weeds. This would make the weeds resistant to the herbicide and therefore much more difficult to control, and possibly result in increased use of herbicides that would kill non-target plants in the surroundings.

The specification states that 'Candidates should be able to evaluate the ethical, social and economic issues involved in the use of genetic engineering in medicine and in food production'. So you should have views on the issues raised by genetic engineering and be able to justify them. There are no 'correct' answers to this type of question — the important thing is to recognise the particular issues in a given situation and to comment on them.

Questions
&
Answers

In this section of the guide there are 14 questions based on the topic areas outlined in the Content Guidance section. The section is structured as follows:

- sample questions in the style of the module
- example candidate responses mostly at the C/D boundary, though sometimes as low as grade E (Candidate A) — these answers demonstrate some strengths but many weaknesses, with potential for improvement
- example candidate responses at the A/B boundary (Candidate B) — these answers demonstrate thorough knowledge, a good understanding and an ability to deal with the data that are presented in the questions. There is, however, still some room for improvement

Some parts of the questions simply ask you to recall basic facts. Other parts contain material with which you are unfamiliar. Before answering these, ask yourself 'Which biological principle is this addressing?' Write down the principle (in rough) and then work out how the principle applies to the data. In calculations, always show your working, and when reading graphs, always draw lines between the plot in question and the axes.

Examiner's comments

All candidate responses are followed by examiner's comments. These are preceded by the icon *e* and indicate where credit is due. In the weaker answers, they also point out areas for improvement, specific problems and common errors such as lack of clarity, weak or non-existent development, irrelevance, misinterpretation of the question and mistaken meanings of words.

The genetic code (1)

The diagram shows part of a **DNA** molecule.

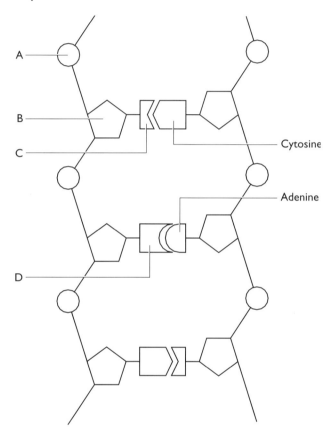

(a) **Name the molecules A, B, C and D.** (4 marks)
(b) **Describe how a DNA molecule replicates itself.** (6 marks)

Total: 10 marks

■ ■ ■

Answer to question 1: candidate A

(a) A: phosphate
 B: sugar
 C: base
 D: thymine

ℯ A is correct. Although B is a sugar, it is not sufficient just to state 'sugar'. RNA and
 DNA have different sugars, so you must state which one. For C it is not sufficient

just to state 'base' — there are four different bases. You will usually be given two of the names and will be expected to identify the other two using the principle of complementary base pairing. D is correct.

(b) The DNA molecule unzips. Free nucleotides then become attached to their complementary bases on DNA chains. The nucleotides then join together to form the two new DNA chains.

    e This answer would obtain 3 of the 6 marks available. Although the main stages are given, there is no reference to either of the enzymes involved. You should always give examples of complementary base pairing to show that you understand the principles involved.

■ ■ ■

Answer to question 1: candidate B

(a) A: phosphoric acid
B: deoxyribose
C: guanine
D: thymine

    e A perfect answer for 4 marks.

(b) To replicate, the DNA molecule has first to unzip. This is brought about by enzymes, which break the hydrogen bonds that hold the two chains of the DNA molecule together. There are four types of free nucleotide available. These become attached to the complementary bases on each of the DNA chains, e.g. adenine nucleotides become attached to thymine bases on the chain and cytosine nucleotides become attached to guanine bases on the chain. The enzyme DNA polymerase then attaches the nucleotides together to form the two complementary DNA chains. Two new DNA molecules are thus formed.

    e A good answer for 6 marks which addresses fully the main stages — 'unzipping', free nucleotides, complementary base pairing and the linking of nucleotides to form chains. Giving examples of complementary base pairing demonstrates an understanding of the principles involved.

The genetic code (II)

Relate the structure of each of the following to its function:
(a) **DNA** (5 marks)
(b) **transfer RNA** (3 marks)

Total: 8 marks

■ ■ ■

Answer to question 2: candidate A

(a) The DNA molecule consists of two chains. Each chain contains four nucleotides. Each nucleotide consists of a phosphate group, a sugar and a base. There are four different bases — adenine, guanine, cytosine and thymine. Adenine in one chain always pairs with thymine in the opposite chain. Cytosine always pairs with guanine. The DNA molecule has many genes, so it stores large amounts of information. It is a very stable molecule.

> *e* There are several weaknesses in this answer which is awarded just 2 marks. In the description of structure it infers that the chain is only four nucleotides long. There is no reference to the hydrogen bonding that holds the chains together and thus provides stability. The section on information is weak. A common error is not stating that a gene is a sequence of nucleotides in triplets called codons. Although the candidate realises that the molecule is stable, there is no explanation as to why this is advantageous.

(b) A tRNA molecule has a clover-leaf shape. One end of the molecule is attached to an amino acid. The other end attaches to mRNA during protein synthesis.

> *e* This answer misses the main point about tRNA molecules — that there are 20 different types, each with a different attachment site for a specific amino acid and each with a different anticodon. It receives just 1 mark.

■ ■ ■

Answer to question 2: candidate B

(a) The DNA molecule consists of a double helix of two nucleotide chains. The nucleotide chains contain four different types of nucleotide. Each nucleotide consists of a phosphate group, a sugar called deoxyribose and a base containing nitrogen. There are four different bases — adenine, guanine, cytosine and thymine. Adenine in one chain always pairs with thymine in the opposite chain, as does cytosine with guanine. The two chains are held together by hydrogen bonds between the bases.

DNA stores information in the form of codons. Each codon consists of three nucleotide bases. DNA is a very long molecule, which means that large amounts

question

of information can be stored on it. Hydrogen bonding makes DNA a very stable molecule.

e A good answer for 4 marks, which only missed one major point — the principle that DNA molecules can replicate themselves and in doing so make exact copies of the genetic information they hold. The last point about stability of the molecules could have been developed further to state why a stable molecule is important — so that the genetic information remains intact and is not easily corrupted.

(b) Transfer RNA is a single-chained RNA molecule composed of RNA nucleotides. The molecule is folded to form a clover-leaf shape. This shape is maintained by hydrogen bonds. At one end of the molecule it has an attachment site for a specific amino acid. There are 20 different tRNA molecules, each with an attachment site for a different amino acid. At the other end it has an anticodon. The anticodon allows it to attach to the complementary codon on mRNA during protein synthesis. Each type of tRNA molecule has a different anticodon.

e A good answer for full marks, which relates the parts of the tRNA molecule to their functions.

The genetic code (III)

(a) **Explain what is meant by semi-conservative replication of DNA.** (2 marks)
(b) **In 1958, Meselson and Stahl published the results of an experiment which supported the theory of semi-conservative DNA replication. Their experiment is shown in the diagram.**

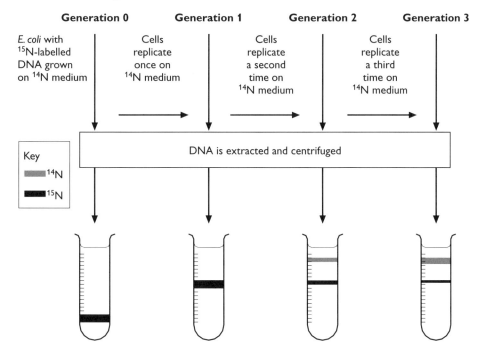

The ^{15}N isotope is heavier than the normal ^{14}N isotope.
(i) **Explain how the results support the semi-conservative theory of DNA replication.** (4 marks)
(ii) **Which parts of the DNA molecule are labelled with ^{15}N?** (1 mark)

Total: 7 marks

■ ■ ■

Answer to question 3: candidate A

(a) Half of each new molecule comes from the original molecule.

    e This answer only goes half way — it states the principle but does not explain it, so only 1 mark can be awarded.

(b) (i) In generation 0, DNA contained only ^{15}N. These are the heaviest so they are at the bottom. In generation 2, some of the DNA molecules would contain only ^{14}N. These molecules are lighter, so they form a band higher up the tube.

(ii) The bases

💬 The answer for part (i) would receive 2 marks — for recognising that some molecules contain only ^{14}N, and that molecules containing only ^{14}N are lighter. However, it does not answer the question which asks how the results support semi-conservative replication. To do this, the answer must refer to DNA strands and their replication. Part (ii) is correct.

■ ■ ■

Answer to question 3: candidate B

(a) When the DNA chains split, a new complementary chain forms for each of the original chains. Half of each new molecule therefore comes from the original molecule; it is this half which is said to be conserved.

💬 A good answer which explains the principle, for 2 marks.

(b) (i) In generation 0, both strands of the DNA contained ^{15}N. In generation 1, one strand would be the original strand containing ^{15}N, but the other strand would be ^{14}N, so the molecules would not be as heavy. The heavier DNA molecules are forced further down the tube during centrifugation. In generation 2, half the DNA molecules would contain only ^{14}N because they are formed from the ^{14}N containing strands of the generation 1 DNA molecules, but half the molecules would contain one ^{15}N and one ^{14}N strand. The molecules containing only ^{14}N strands are lighter than the ones containing both ^{14}N and ^{15}N, so they form a band higher up the tube.

(ii) The bases

💬 Part (i) is an excellent answer for 3 marks, which could only have been improved by explaining the difference between the widths of the bands in generations 2, 3 and 4. (The wider the band, the more strands of DNA there are containing that particular isotope.) Part (ii) is correct.

The genetic code (IV)

The diagram shows the sequence of bases of part of one **DNA** strand. The sequence should be read from left to right.

C	A	G	A	C	C	A	C	C	C

(a) (i) Give the base sequence for the part of the mRNA molecule which would be produced by this **DNA** sequence. (2 marks)

(ii) Give the bases of the tRNA molecules that correspond to the shaded area of the sequence. (1 mark)

(b) The table shows some **DNA** base sequences and the amino acid sequences for which they code.

DNA base sequence	Amino acid
CGA	Alanine
CCC	Glycine
CCA	Glycine
GAC	Leucine
ACC	Tryptophan
CAG	Valine

As a result of a mutation, the first adenine base (**A**) from the **DNA** sequence in the diagram is lost.

(i) Use the table to identify the amino acids that the **DNA** sequence would now code for. (2 marks)

(ii) Give two other ways in which gene mutations occur. (2 marks)

Total: 7 marks

■ ■ ■

Answer to question 4: candidate A

(a) (i) G T C G G T G G G

(ii) U G G

🖉 In (i) the candidate has forgotten that in mRNA thymine (T) is replaced by uracil (U) and loses 1 mark. In (ii) the candidate has transcribed the code from DNA rather than translated from mRNA — no marks.

(b) (i) (1) Alanine

(2) Glycine

(3) Tryptophan

(ii) Substitution
Addition

Alanine is correct in (i) but the candidate has not realised that all the other triplets have altered as a result of the deletion. Part (ii) is correct.

■ ■ ■

Answer to question 4: candidate B

(a) (i) G U C G G U G G G
   (ii) A C C

(b) (i) (1) Alanine
      (2) Glycine
      (3) Glycine
  (ii) Substitution
      Addition

All correct, receiving full marks.

The genetic code (V)

(a) Explain how cells use **DNA** to produce proteins. (6 marks)

(b) Explain in terms of protein production why some mutations are harmful while others are harmless. (4 marks)

Total: 10 marks

■ ■ ■

Answer to question 5: candidate A

(a) DNA carries genetic information. The first stage in the protein production is transcription where this genetic information is copied. The DNA molecule unzips and the sense strand acts as a template for the production of a molecule of mRNA. The RNA nucleotides are joined together by RNA polymerase. The mRNA molecule then passes out of the nucleus and becomes attached to a ribosome. There are 20 different types of tRNA molecule. tRNA molecules bring the amino acids to the mRNA on the ribosomes in the correct sequence. This is called translation. The amino acids are then joined together to form a polypeptide chain.

ℓ A fair answer for 4 marks, mentioning both transcription and translation. However, there are no references to codons and anticodons, so this is a description of the process rather than an explanation.

(b) A gene mutation results in a change in the genetic code of DNA. A mutation such as a substitution might affect only a small region of the DNA, resulting in only one different amino acid being assembled into the protein. If the new amino acid does not affect the tertiary structure of the protein it will probably not affect the functioning of the protein.

ℓ Although the candidate has the right idea about the effect of mutation on protein structure, there is no reference to codons, nor the different effects that the three types of gene mutation (addition, deletion and substitution) have on protein production. Only 2 marks are awarded here.

■ ■ ■

Answer to question 5: candidate B

(a) DNA carries genetic information in the form of triplet codons. Each codon is a sequence of three nucleotide bases. The four DNA bases are adenine, thymine, cytosine and guanine. A gene contains all the codons needed to produce the particular protein. The first stage in the protein production is transcription. The DNA molecule unzips and the sense strand acts as a template for the production of a molecule of mRNA. The RNA nucleotides line up with the complementary DNA codons and are then joined together by RNA polymerase.

The mRNA molecule then passes out of the nucleus and becomes attached to a ribosome. There are 20 different types of tRNA molecules. Each of these has a different type of amino acid attached to it at one end. At the other end of the tRNA molecule is a triplet anticodon. The anticodon is complementary to the mRNA code for the particular amino acid.

tRNA molecules bring the amino acids to the mRNA on the ribosomes. The codon–anticodon pairings result in amino acids being brought to the mRNA in the correct sequence. This is called translation. The amino acids are then joined together by condensation reactions to form a polypeptide chain.

> A good answer for full marks, explaining both transcription and translation. The descriptions of the different stages in the process are grouped into paragraphs.

(b) A gene mutation results in a change in the base sequence of DNA. A mutation such as a substitution might affect only one codon, resulting in just one different amino acid being assembled into the protein. If the new amino acid does not affect the tertiary structure of the protein it will probably not affect the functioning of the protein. A gene mutation involving an addition or a deletion of a nucleotide could cause a change to all the subsequent codons. This would affect many different amino acids to be assembled into the protein, resulting in significant changes to its tertiary structure. The protein might then not function.

> A good answer for full marks, explaining the different effects of substitution and addition/deletion.

The cell cycle (1)

The diagram shows stages of mitosis in an animal cell.

A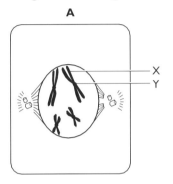

B

C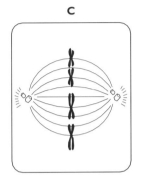

D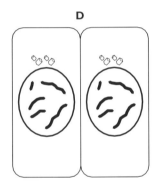

E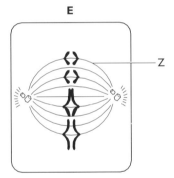

(a) **Which of the drawings shows:**
   (i) anaphase
   (ii) metaphase
   (iii) telophase? (3 marks)
(b) **Name the structures labelled X, Y and Z.** (3 marks)
(c) **Explain why cells produced by mitosis have exactly the same genetic information as the parent cell.** (3 marks)

Total: 9 marks

■ ■ ■

Answer to question 6: candidate A

(a) (i) E
   (ii) C
   (iii) D

(b) X: chromosome
   Y: centromere
   Z: spindle

> *e* The answers to (a) are all correct. In part (b) only Y and Z are correct, although for Z 'spindle fibre' would be more exact. In labelling, make sure you distinguish between chromosome and chromatid.

(c) The two chromatids of each chromosome have identical genetic information. During anaphase the chromatids of each of the chromosomes are pulled to opposite poles of the spindle.

> *e* This is an incomplete explanation and is awarded just 1 mark. In questions such as this, you must always refer to replication of DNA during interphase and the identical DNA molecules possessed by each of the two chromatids that make up a chromosome.

■ ■ ■

Answer to question 6: candidate B

(a) (i) E
   (ii) C
   (iii) D

(b) X: chromatid
   Y: centromere
   Z: spindle fibre

(c) During interphase, DNA replication takes place. The two chromatids of each chromosome receive identical DNA molecules. During anaphase the chromosomes are pulled to opposite poles of the spindle. Each daughter cell receives an identical set of chromatids and therefore an identical set of DNA molecules.

> *e* Perfect answers to (a) and (b) for full marks. In (c) 1 mark was lost by stating that chromosomes, rather than chromatids, move to the poles.

The cell cycle (II)

The diagram shows how carrot plantlets (small, young plants) can be obtained from small pieces of carrot tissue.

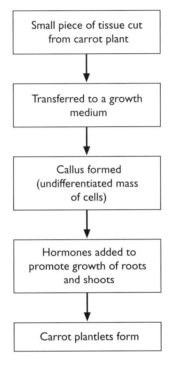

Small piece of tissue cut from carrot plant

↓

Transferred to a growth medium

↓

Callus formed (undifferentiated mass of cells)

↓

Hormones added to promote growth of roots and shoots

↓

Carrot plantlets form

(a) **Explain why small pieces of carrot have the ability to develop into whole carrot plants.** (2 marks)

(b) **Suggest and explain why growers might wish to produce carrots in this way rather than growing them from seeds.** (3 marks)

Total: 5 marks

■ ■ ■

Answer to question 7: candidate A

(a) Every cell must be capable of becoming a root or a stem cell so they must all have the same genetic information.

> *e* Only a partial explanation for 1 mark — there is no mention of why all the cells of the carrot have identical genetic information.

(b) The grower had bred a variety of carrot that he liked and wanted to make clones from it.

e Only part of the explanation, so only 1 mark is awarded. There is no reference to sexual reproduction producing seeds with resulting variation. Always try to offer specific scientific suggestions of qualities rather than 'variety of carrot that he liked'.

■ ■ ■

Answer to question 7: candidate B

(a) Every cell in a carrot plant has an identical copy of the genetic information needed to produce a carrot. This is because they are all produced from the zygote by mitosis, during which the carrot DNA is replicated. The differentiation of these cells into stems and roots, therefore, only needs the application of hormones.

e A good answer, in terms of mitosis and DNA replication, for full marks. Always try to include explanations for information included in the question, in this case the role of the hormones.

(b) Seeds are produced by sexual reproduction, which involves mixing the genetic information of two different plants via their gametes. Producing plantlets is a form of vegetative propagation which involves only mitosis. The breeder may have developed, for instance, a disease-resistant variety of carrot and wish to produce clones of it.

e A good answer, in terms of comparing mitosis and cloning with sexual reproduction, for 2 marks. It would have been improved by stating that the desired characteristic might not appear in the phenotype of the sexually produced organism. Be prepared to offer suggestions as to why clones might be desirable — in this case to pass on disease resistance.

The cell cycle (III)

The diagram shows a buttercup plant.

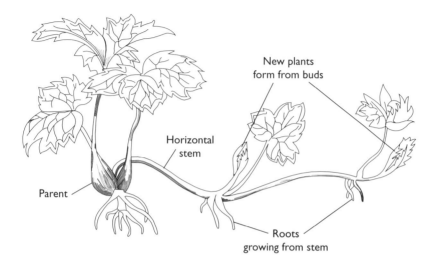

The buttercup spreads by producing horizontal stems, which eventually produce new plants.

(a) The buttercup plants produced by the method shown in the diagram are identical to the parent plant. Explain why. (3 marks)

(b) Describe how several identical sheep could be obtained from one pair of parents. (5 marks)

Total: 8 marks

■ ■ ■

Answer to question 8: candidate A

(a) Because the cells have been produced by asexual reproduction, no other buttercup was involved.

> *e* This answer receives only 1 mark, for realising that asexual reproduction is involved. There is no reference to the replication of genetic information, or to the role of mitosis in ensuring that each daughter cell receives an identical copy of the information.

(b) The fertilised egg is allowed to divide to the 16-cell stage, then the cells are separated. Each cell will divide to produce an identical sheep.

> *e* This answer receives only 2 marks — for stating that a fertilised egg is used and that it is allowed to divide several times before it is split into separate cells. There are no references to fertilisation or to further development of the egg.

■ ■ ■

Answer to question 8: candidate B

(a) The plants are produced by asexual rather than by sexual reproduction. Asexual reproduction involves cells dividing by mitosis. In mitosis the DNA replicates during interphase, forming two identical chromatids. Identical chromatids then go to opposite poles of the spindle during anaphase, ensuring that each daughter cell receives the same genetic information.

e An excellent answer for full marks.

(b) Eggs cells are taken from a female sheep and fertilised. The fertilised egg is then allowed to divide until it has formed a ball of cells. The ball is divided into individual cells, and each cell is allowed to divide to form a ball of cells. Each ball of cells is then inserted into a female sheep to develop.

e A good answer for 4 marks which outlines the main stages in the process. Full marks would have been obtained for further detail, such as in vitro fertilisation or implantation of the embryo into the uterus of the female.

Sexual reproduction and fertilisation (I)

The diagram shows the life cycle of a single-celled organism.

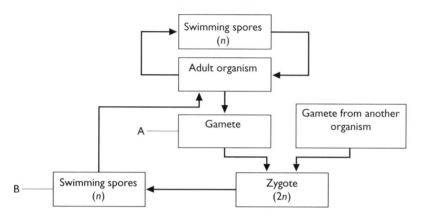

(a) Complete the boxes to show which stages are diploid (2n) and which are haploid (n).   (2 marks)

(b) On the diagram, mark with an X one place where mitosis occurs and mark with a Y where meiosis takes place.   (2 marks)

(c) Will the stages A and B have the same genetic information? Give the reason for your answer.   (2 marks)

Total: 6 marks

■ ■ ■

Answer to question 9: candidate A

(a) and (b)

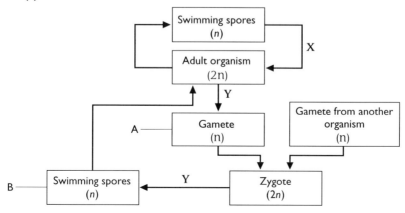

question

e For (a), the candidate made the common mistake of assuming that all adult organisms are diploid. Since the adults are formed from haploid swimming spores, they must also be haploid. For (b), since the candidate assumed the adult was diploid, the meiosis box is incorrect.

(c) They have different genetic information because the genetic information from A had mixed with that of another gamete to form the zygote from which B is formed.

e This answer has the correct basic idea for 1 mark, but always try to use correct terminology such as fuse and fertilisation.

■ ■ ■

Answer to question 9: candidate B

(a) and **(b)**

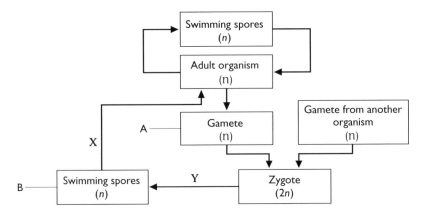

e For both (a) and (b) the diagram was completed correctly.

(c) They would have different genetic information because the gamete, A, fuses with a gamete from another organism during fertilisation. The zygote will therefore contain genetic information from both gametes. Swimming spore B is produced from the zygote and therefore will contain some genetic information from both of the gametes.

e A good answer for full marks, which appreciates that fertilisation results in a mixing of genetic information.

Sexual reproduction and fertilisation (II)

Greenfly are small insects that live on many species of flowering plant. The diagram shows the life history of one species of greenfly.

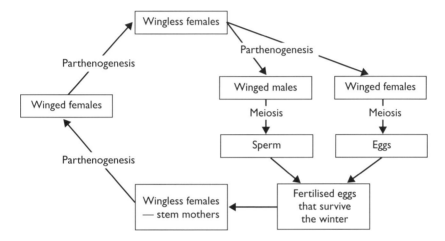

Wingless females, called stem mothers, produce living young without fertilisation throughout the summer. This process is known as parthenogenesis. Eventually the plant host of the stem mother and her offspring becomes overcrowded; when this occurs, some aphids develop two pairs of wings and fly off to new plants. In late summer, both males and females are produced by parthenogenesis. After they mate, the female lays fertilised eggs that survive the winter. In spring, the fertilised eggs hatch to form stem mothers.

(a) State whether each of the following is haploid or diploid. Give the reason for your answer in each case.
  (i) Stem mother
  (ii) Sperm
  (iii) Winged female   (3 marks)
(b) Will the stem mothers produced in one year be genetically identical to the stem mothers produced in the following year? Give the reason for your answer.   (2 marks)

Total: 5 marks

■ ■ ■

Answer to question 10: candidate A

(a) (i) Diploid
   (ii) Haploid
   (iii) Diploid

e All three answers are correct, but the candidate receives only **1** mark because there are no explanations for the answers.

(b) She would not be genetically identical because she is produced by sexual reproduction.

e This answer receives only **1** mark since there is no explanation for why sexual reproduction produces genetic variation.

■ ■ ■

Answer to question 10: candidate B

(a) (i) The stem mother is diploid because she develops from a fertilised egg.
(ii) The sperm is haploid because it is formed by meiosis.
(iii) The wingless mother is haploid because she is produced without fertilisation.

e (i) and (ii) are correct for 2 marks, but the correct answer to (iii) is diploid, since only diploid organisms produce gametes by meiosis.

(b) Since they are produced by sexual reproduction, they might receive different alleles from each parent, and therefore might not be genetically identical.

e A good answer for full marks.

Applications of gene technology (I)

A herbicide called glyphosate works by blocking the activity of an enzyme that most plants need in order to synthesise amino acids. A mutant petunia plant has been found which contains a gene that gives resistance to glyphosate. Scientists decide to transfer this gene to crop plants.

(a) Suggest how the mutant gene could be transferred from petunia plants into a crop plant and how the transformed crop plant could be propagated. **(6 marks)**

(b) Explain how farmers would benefit from growing the transformed crop plants. **(2 marks)**

(c) Describe one environmental risk associated with introducing this mutant gene into crop plants. **(2 marks)**

Total: 10 marks

■ ■ ■

Answer to question 11: candidate A

(a) It could be cut from the plant using a restriction endonuclease enzyme. A ligase enzyme can be used to insert the mutant gene into plasmids. The plasmids could then be inserted into the crop plant.

Ⓔ Although the candidate mentions both enzymes correctly, there is lack of detail required to gain high marks. There is no reference to the gene as a section of DNA, to using the same endonuclease to cut the plasmid or to multiplying the plasmids in bacteria. The last part of the question is almost totally ignored. Only 2 marks can be awarded here.

(b) The crop could be sprayed with herbicide which would kill the weeds but not affect the crop.

Ⓔ Only half an answer, so 1 mark — there is no reference to how this would benefit the farmer.

(c) The herbicide-resistance gene might transfer from the crop plant to weed plants. The weeds would then grow out of control.

Ⓔ The answer starts with the correct idea for 1 mark, but there is no suggestion as to how the gene might transfer to the weeds. At AS level, the statement 'would then grow out of control' is rather weak.

■ ■ ■

Answer to question 11: candidate B

(a) The section of DNA containing the mutant gene can be cut from the DNA of the petunia plant using a suitable restriction endonuclease enzyme. Plasmids from a bacterium can then be cut open with the same endonuclease enzyme. A suitable

ligase enzyme can be used to insert the mutant gene into the plasmids. The plasmids can be inserted into bacteria which then reproduce to produce many transformed plasmids. The transformed plasmids could be extracted from the bacteria and then injected into cells from a young embryo of the crop plant. The transformed crop plant cells could be grown into plantlets on a suitable growing medium.

🖉 A good answer for 5 marks, which refers both to the genetic engineering techniques and to propagation. Full marks could have been obtained by reference to marker genes or to selection.

(b) Herbicides are used to kill weeds that grow amongst crops. The transformed crop could be sprayed with herbicide which would kill the weeds but not affect the crop. The farmer would get a greater yield.

🖉 A good answer for 2 marks which refers both to the effect on the plants and to the benefit for the farmer.

(c) The herbicide-resistance gene might transfer from the crop plant to weed plants via pollen. If the weeds became resistant to the herbicide, some other method would have to be found to kill them.

🖉 A good answer for 2 marks which gives both a mechanism and a consequence in the explanation.

Applications of gene technology (II)

The corn borer is an insect pest that affects maize plants. Scientists have isolated a gene that produces an insecticide gene which can kill the corn borer. The technique shown in the flow chart below was used to introduce the gene into maize cells. A gene for resistance to the antibiotic ampicillin and a gene for resistance to weed killer were also inserted into the plasmids.

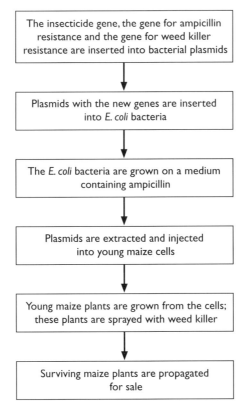

The insecticide gene, the gene for ampicillin resistance and the gene for weed killer resistance are inserted into bacterial plasmids

↓

Plasmids with the new genes are inserted into *E. coli* bacteria

↓

The *E. coli* bacteria are grown on a medium containing ampicillin

↓

Plasmids are extracted and injected into young maize cells

↓

Young maize plants are grown from the cells; these plants are sprayed with weed killer

↓

Surviving maize plants are propagated for sale

(a) Explain why the ampicillin-resistance gene was added to the plasmids. (3 marks)
(b) Explain why the weed killer-resistance gene was added to the plasmids. (3 marks)
(c) Explain the possible advantages and drawbacks of introducing insecticide-producing genes into plants. (5 marks)

Total: 11 marks

■ ■ ■

Answer to question 12: candidate A

(a) When the bacteria are grown on the medium containing ampicillin, only the bacteria that have the ampicillin-resistance gene will grow.

This is correct as far as it goes, for 1 mark, but there is no reference to these bacteria containing the other two genes.

(b) Spraying with weed killer will kill the maize plants that do not have the weed killer-resistance gene. The surviving plants will be able to produce insecticide.

 This is also true as far as it goes, for 2 marks, but the explanation does not state *why* this stage is necessary (to select only those plants that have weed killer resistance).

(c) The plant would produce insecticide that would kill any corn borers that landed on the plant. The corn would grow better, so the farmer would make more money. A drawback is that friendly insects could also be killed.

 This answer receives 2 marks. The candidate has identified an advantage and a drawback, but has used 'everyday' language, such as 'grow better' and 'friendly insects', rather than scientific terminology.

■ ■ ■

Answer to question 12: candidate B

(a) The ampicillin-resistance gene is a marker gene. When introducing genes into plasmids, not all the plasmids will have the genes successfully introduced. When the *E. coli* bacteria are grown on the medium containing ampicillin, only the bacteria with transformed plasmids will grow. This means that only plasmids containing the new genes will be used to transform the maize plants.

 A good answer, for full marks, showing sound understanding of the use of antibiotic-resistance marker genes.

(b) To be successful, the DNA from the transformed plasmids must join with the DNA in the maize cells. This will only happen in a small proportion of the maize cells. Spraying with weed killer will kill the maize plants where this did not happen. The surviving plants will contain the weed killer-resistance gene.

 A sound explanation for 2 marks. This could have been improved by stating that the surviving plants contain *both* desired genes.

(c) Corn borers are insect pests that consume some of the materials that the corn plant produces by photosynthesis. This will reduce the yield of the crop, so the farmer will make less profit. The genetically engineered crop will produce its own insecticide. Also, because insecticides do not have to be sprayed, less insecticide will enter the soil or water supplies. The insecticide might kill non-target species in addition to the corn borers.

 A good answer which could only have been improved by reference either to the effect of insecticide on food webs, or to comparative costs of using insecticide-producing corn rather than spraying insecticide. Nevertheless, full marks are awarded.

Applications of gene technology (III)

Gene probes are used in screening for genetic diseases. A gene probe is a single-stranded piece of DNA labelled with a radioactive isotope. The diagram shows how gene probes are used in diagnosis.

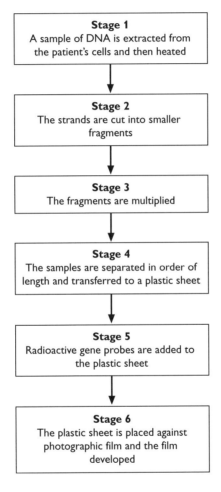

Stage 1
A sample of DNA is extracted from the patient's cells and then heated

Stage 2
The strands are cut into smaller fragments

Stage 3
The fragments are multiplied

Stage 4
The samples are separated in order of length and transferred to a plastic sheet

Stage 5
Radioactive gene probes are added to the plastic sheet

Stage 6
The plastic sheet is placed against photographic film and the film developed

(a) Why is the DNA sample heated in stage 1?　　　　　　　　　　　　　　(1 mark)
(b) Name the enzyme used to cut DNA into smaller fragments.　　　　　　(1 mark)
(c) Name the process used to multiply the fragments in stage 3.　　　　　(1 mark)
(d) Name the process used to separate the DNA fragments in stage 4.　　(1 mark)
(e) Explain how the radioactive probes would detect whether some of the fragments contain the disease gene (stages 5 and 6).　　　　　　　　　　(3 marks)

Total: 7 marks

■ ■ ■

Answer to question 13: candidate A

(a) To split the DNA
(b) Endonuclease
(c) PCR
(d) Electrophoresis
(e) The probe would show up when the film is developed

> The candidate has not given quite enough information in (a) — the two chains should have been mentioned. The correct answers are given for (b) and (c), but always try to give names rather than abbreviations. The correct answer is given for (d). In (e) the candidate has not stated that the probe attaches to the disease gene, nor explained why it does this — it has a complementary base sequence.

■ ■ ■

Answers to question 13: candidate B

(a) To split the DNA into its two chains
(b) Restriction endonuclease
(c) Polymerase chain reaction
(d) Electrophoresis
(e) The probe would attach to the disease gene. The band with the disease gene plus the marker would then show up when the film is developed.

> Answers (a) to (d) are correct. Part (e) is correct as far as it goes, but loses 1 mark as the candidate should have explained that the probe attaches to the disease gene because it has a complementary base sequence.

Applications of gene technology (IV)

Cystic fibrosis is the result of an altered protein.
(a) (i) Name the protein that works normally in healthy people but does not
work in people with cystic fibrosis. (1 mark)
(ii) Explain how the altered protein is produced. (4 marks)
(iii) Explain how the altered protein accounts for the symptoms of cystic fibrosis. (4 marks)
(b) Describe one way in which gene therapy is used to treat cystic fibrosis. (3 marks)

Total: 12 marks

■ ■ ■

Answer to question 14: candidate A

(a) (i) CFTR
(ii) The defective protein is caused by a mutation. CFTR has one missing amino acid.
(iii) The defective CFTR is unable to transport chloride ions across the membranes of the lungs. The mucus outside the cells becomes thicker. It is more difficult for oxygen to diffuse into the blood as quickly, so the person gets out of breath.

e Part (i) is correct. In part (ii) both statements are correct, for 2 marks, but a good explanation must link the cause with the effect. The answer for (iii) makes three points reasonably well, for 3 marks, but does not link chloride ions with water accumulation and thickening of the mucus.

(b) The healthy gene is wrapped in lipid. Because there is lipid in the membrane, the gene will be delivered to the membrane.

e This receives 1 mark for a correct method of delivering the gene, but there is a fundamental error — the gene is delivered to the cells which then use the genetic information to produce membrane protein.

■ ■ ■

Answer to question 14: candidate B

(a) (i) CFTR.
(ii) The defective protein arises from a mutation in the gene for CFTR production. This mutation is caused by the deletion of one of the codons in the gene so that defective CFTR has one missing amino acid.
(iii) The defective CFTR has a different tertiary structure because of the missing amino acid. Due to the change in its shape it is unable to transport chloride ions across the membranes of the lungs. The cells therefore retain water because the chloride ions give a more negative water potential, and the mucus outside the

cells becomes thicker. This makes it more difficult for oxygen to diffuse into the blood. The person will very easily be out of breath.

✎ Part (i) is correct. A good answer, for 3 marks, has been given for part (ii) which refers both to mutation and to the effect that this has on production of protein. Another good answer has been given for part (iii), which links the tertiary structure of the altered protein to both the functioning of the protein and the disease symptoms, and is awarded full marks.

(b) A virus that can infect human lung cells is made harmless by removing the genes that damage human cells. The healthy CFTR gene is then inserted into the virus DNA. The virus is delivered in an aerosol. The virus has proteins in its coat that can penetrate the membrane of the lung cells. When the virus DNA is inserted into the lung cells, the healthy gene is also inserted.

✎ A good answer, for full marks.